Barrow Steelworks
Continuous Casting
1951 - 1961

S. Henderson and K. E. Royall

Published in 2021 by Stan Henderson
© Copyright Stan Henderson

ISBN: 978-1-913898-24-3

Book & Cover Design by Russell Holden

 Pixel Tweaks Publications
SELF PUBLISHING MADE SIMPLE

www.pixeltweakspublications.com

A Catalogue record for this book is available from the British Library.

Printed by Ingram

Definition of Steel

Steel is an alloy of iron, carbon and other elements as required, which is refined while in the fully molten condition and brought to final analysis while fully molten.

Hephaestus: Greek god of metal working *image supplied by author*

FOREWORD

Continuous Casting is fundamentally simple. Seen in action it is deceptively so. Molten steel is poured at a controlled rate into a bottomless, water-cooled, mould. From the bottom of the mould emerges a continuous strand of billet, bloom or slab to be cut automatically into the lengths required for further processing. There are few operatives to be seen about the plant and the whole process goes like the proverbial clockwork. Notwithstanding, there was a hundred years between the patented scheme of Bessemer, and a viable method finally emerging. The process has given the steel industry a valuable, economic, production method, dubbed – A Technical Revolution in Steelmaking.

This book is an attempt to recount at least part of the story.

IT HAD ALWAYS BEEN THE
DREAM OF STEELMAKERS
THAT STEEL COULD BE MADE
IN A TRULY CONTINUOUS
PROCESS. DURING THE 1950S,
IAIN HALLIDAY AND HIS
COLLEAGUES AT BARROW
MOVED THAT DREAM CLOSER
TO REALITY.

Two photos of the process in action during 1958. Top image shows the oil-fired ladle teeming liquid metal into a transverse launder and thence into two separate tundishes. The bottom image is of the mould deck where two types of control can be seen. The operator is controlling one mould manually, while the other is operating under remote control. S. Henderson collection

- Preamble -

Early in 2017 the authors were invited to contribute to a collective work on the development and subsequent success of the Continuous Casting of Steel. The work, which was to comprise of accounts from each of the countries having an involvement, would be as truly international as the process has been in origin and application. The resultant work, *Stranggießen–Continuous Casting* (Aschendorff Verlag), (hereinafter referred to as Ref. 1) was planned, coordinated and compiled by Professor Doctor, Manfred Rasch, Chief Archivist of the *ThyssenKrupp* organization of Germany. It was published in 2018.

Printed mainly in German, the book is a specialist encyclopaedia aimed, chiefly, at the *cognoscenti*. Accordingly, for the purpose of providing an accessible local record of Barrow's contribution to the process, described as a Technical Revolution in the Steel Industry, the authors have decanted the Barrow Chapter into this current work – adding more of the local application that space restriction precluded the inclusion of in the original book. The work you have before you originated from various sources, namely – *Continuous Casting at Barrow* by Iain M. D. Halliday; Ken Royall's involvement as manager of the Fuel and Instrument Department of Barrow Steelworks; research undertaken on the 2015 book – *Barrow Steelworks, an Illustrated History*; interviews with plant operatives, technical staff and service tradesmen.

Although not originally conceived in Barrow, the continuous casting of steel, in twenty years, had not advanced significantly beyond the experimental stage. What Barrow works achieved – during six years of intense development work – was to progress the concept into a commercially viable process.

The main stumbling block – and there were several - that at the time seemed insurmountable, was the tendency of the liquid metal to stick to the internal surface of the copper mould despite the use of mould lubricants. This then, together with breakouts (discussed later) frustrated most attempts to completely drain a charge of molten metal through the casting machines.

The problem was eventually resolved at Barrow, as attested by Roderick Guthrie and Mihaiela Isac in their essay – *The Development of Continuous Casting Machines for the production of Steel in North America*, reprinted from ref. 1 and reproduced below:-

Iain Halliday and his colleagues at Barrow-in-Furness, UK resolved the key resolution to the problem of molten steel freezing, and then sticking on to the walls of the copper mould. They developed the 'Negative Strip' concept' in which the down-stroke of the oscillating mould was adjusted to be slightly faster than the withdrawal speed of the billet being cast. This solved the problem of steel sticking and at the same time, lead to the healing of any tears in the forming steel skin of the billet. That, in turn, led to the near total elimination of breakouts. With that, CC machines for steel were here to stay, and gradually usurped ingot casting practices around the world. It is worth noting that Halliday established a world record by casting at 14.5m/minute without any breakouts, a record that stands to this day! Nowadays, the 1.6 billion tonnes of liquid steel being cast into semi-finished products all around the world, uses this casting paradigm.

Although much of the credit for developing the process must go to the Germans and the West German concern - *Mannesmann* - it is perhaps fitting that English inventor, Sir Henry Bessemer started the ball rolling as far back as 1856. His idea being to pour liquid metal between two, rotating, water-cooled rolls so as to solidify it into a thin rolled strip or sheet. The idea wasn't followed-up, for one thing the pre-requisites, such as the industrial gases - oxygen and propane - weren't available in bulk during the late 19th century. In the intervening hundred years many different ideas for the direct pouring of liquid metal continuously to semi-finished products were suggested. In spite of the many patents registered, progress with steel was negligible and it was in the non-ferrous field that the first continuous casting machines were first developed, particularly for copper, brass and aluminium, where the high quality of the product constituted the principal attraction.

This book, which is the third in the *Barrow Steelworks* series, covers the period between 1951-1961 and includes a description of the works at Barrow, post WW2, as well as key personnel and those actively involved, some hitherto uncredited, in the successful development of the high-speed process.

An appendix is included which contains several images of the Barrow Pilot Plant; Barrow personnel; several photos including some of certain engineering aspects as well as views of the northern-end of the Works.

S. Henderson, 2021

Kirkby-in-Furness, UK

Contents

- THE PIONEERS -

There have been many pioneers of continuous casting – too many to list here. The following are those whose work was relevant to Barrow.

Sir Henry Bessemer

English inventor, (1813 – 1898) was the son of an engineer. He registered over 100 patent applications in his lifetime. Most notable being for his invention of an inexpensive process of mass producing mild steel from iron (1855). Knighted in 1879, he started the ball rolling with regards to the continuous casting of steel.

Sir Henry Bessemer

Siegfried Junghans

German engineer and metallurgist, (1887-1954), had been the technical and commercial manager of a brass foundry, Schwarz-wald AG. He later became CEO of the Black Forest Metal Trading AG, Company. During the inter-war years he began experimenting with the casting of non-ferrous metals using an open-ended mould.

Upon encountering problems with the molten metal sticking to the internal surface of his mould, he devised and patented a non-harmonic mould oscillation technique. This was around 1933. Then, during 1935 – 36, he began applying his System to the continuous casting of steel, this was at Schorndorf, West Germany.

Irving Rossi (1889 – 1991)

American banker, financier, 'engineer' and visionary, was working in Europe as an agent for an investment banking house when he met with Junghans. This was around 1936. Rossi immediately saw the potential if the Junghan's system could be applied to steel. Upon acquiring the exclusive rights to Junghan's patent, he agreed to finance the commercialisation of the process outside of Germany. With the outbreak of war progress became frustrated. After the war Rossi established a relationship with Consul Knut Edstrand, one of the three shareholders in Broderna Edstrand

AB – a steel trading company in Sweden. This was in 1951. Edstrand, being fascinated by the new casting technology, accepted the task of supplying a continuous casting machine, based on the Junghans system, to Barrow Steelworks. In 1954 Rossi met with his future partner Heinrich Tanner and together they founded Concast AG.[1] This company would go on to make casting machines as well as other items

Irving Rossi snapped in his 90th year
(Ref. 1)

of steelworks plant, becoming a world-wide success story. (During 2012 Irving Rossi was inducted into the American Metal Market's Hall of Fame).

Iain Macdonald Dewar Halliday (1918 – 1976)

Graduated from the University of Glasgow in 1940. He went on to be Head of Continuous Casting Research and Consultant on Continuous Casting for the United Steel Companies Ltd.

---

1 Note that Concast AG was founded after Negative Strip had become established at Barrow.

After the end of World War II Halliday came to Barrow where he was, initially, active in the modernisation of Barrow's open hearth plant. From 1950 he was appointed to run with the development of continuous casting, becoming involved in negotiations with Irving Rossi and the subsequent 'Rossi Agreements'. Colleagues would describe him as a 'driven' individual, beset of a determination to succeed (sometimes the lights in his office would burn until midnight).

During 1951 he set-up and chaired the Continuous Casting Steering Committee comprising United Steel and Barrow works personnel.

He lived with his family at Croslands Park, Barrow. He has, in the opinion of the authors, been unfairly over-looked by the UK honours system.

Iain Halliday photographed in his office at Barrow Steelworks in 1962. Courtesy of K.E. Royall

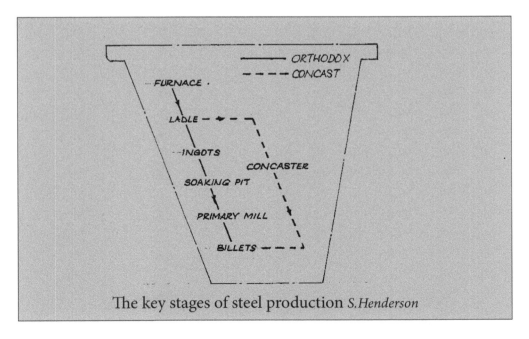

The key stages of steel production *S. Henderson*

- Barrow Works, Post WWII -

During the latter part of the nineteenth century, when Queen Victoria reigned over almost one third of the globe, Barrow-in-Furness was at the centre of the steel-making universe. Then, at the end of World War II, twenty years after the last Bessemers had been blown-out, the works at Hindpool were just a huge, Victorian relic!

The offices and main entrance on Walney Road. Courtesy of K. E. Royall.

By the nineteen fifties things were beginning to change in the world (The steel industry had been nationalised in 1949 by the Labour government – and then reversed by the incoming Conservatives in 1951). In May 1950 the French Foreign Minister announced that the French and German governments were going to create a common *High Authority* to regulate and control the production of coal and steel. In 1952 the Treaty of Paris created the European Coal and Steel Community made up of France; West Germany; Italy; Holland; Belgium and Luxembourg. Britain chose not to join this club.

During the war, the steelmaking side of the works had been requisitioned by the Government and became Ministry of Supply – Barrow Works. The United Steel Companies of Sheffield were appointed managing agents [see diagram on p.8]. From 1942 orders for rails and track products were diverted up the coast to Workington – which had been under the control of United Steel since 1919. Barrow's core business, going forward, would be making the steel billets and slabs required by the re-rolling mills of its hoop and bar mills.

Upon taking over the running of the works, United Steel appointed its own man, Colonel G.N.F. Wingate OBE (1908-1986), to manager. [George Nigel Fancourt Wingate, a no-nonsense character, was the younger brother of Major General Orde Charles Wingate (1903-1944) a senior British Army officer remembered for his unconventional military thinking and believer in the value of surprise attacks!].

The redoubtable Colonel Wingate, new Steelworks manager. Courtesy of K. E. Royall.

William Killingbeck, the last general manager of the entire Hindpool complex, accepted a side-ways move, focusing his efforts on the now separate Ironworks. William Killingbeck, chairman and managing director of Barrow Ironworks Ltd. Mr Killingbeck, a Blackburn man, was the last to manage the entire complex.

WHY is the success story so popular ? What do people find so interesting about the rise of some ordinary chap to a position of power and influence ?

Perhaps it is self-identification with the story's subject— the inspiration of yet another example that even in today's increasingly specialised world, such feats can still be accomplished.

Such a story is that of Mr. William Killingbeck, who died recently while on a business trip to London. Mr. Killingbeck, a Blackburn-born clerk rose to become chairman and managing director of Barrow Ironworks Ltd and gained the reputation of being one of the most enlightened industrialists in the country.

Humble beginning

He arrived at Barrow shipyard as a timekeeper during the latter part of the first world war and became a secretarial clerk at the old Barrow Hematite Steel Company, in 1919.

Despite the energies he put into his work, Mr. Killingbeck still had plenty of drive left over for other interests. He became an active worker at Emmanuel Congregational Church in Abbey-road, Barrow, where he was organist for a while, treasurer and took an active interest in the Sunday school.

In his younger days he had been a keen boxer and footballer and maintained his love of sport, which he supported generously, all his life.

The iron trade, despite all the safeguards cannot be regarded as other than hazardous, and Mr. Killingbeck was extremely keen on his works first-aid system. He was also president of the Barrow Centre of the St John Ambulance Association and was appointed an Officer Brother of the Order.

Hard to replace

The esteem of his employees is characterised in a letter by Mr. F. Cartmel, secretary of the Barrow

*T*HE late MR. WILLIAM KILLINGBECK, chairman and managing director of the Barrow Ironworks. Ltd.

Lodge of the National Union o Blastfurnacemen. Ore Miner: Coke Workers and Kindred Trade: who wrote of the consideration h gave to his men and added " ; good employer is very hard t replace."

The same qualities of fairnes and consideration marked hi work on the Bench, of whic. he became a rota chairman.

Latterly, he became intereste in farming and socially he was popular figure in the Furnes district. He was president of th Barrow branch of the Roy: Society of St George.

Taken all round, he will b greatly missed in the Barrow are:

Photo: N.W. Evening Mail courtesy of Susan Benson (Cumbria Archives & Local Studies Centre – Barrow).

He gained the reputation of being one of the most enlightenened industrialists in the country. Barrow's 60 years old rail mill, driven by 3 sets of horizontal reversing steam engines, that collectively developed 12,000 hp., was adapted for rolling the 2-ton ingots (produced by the open hearth plant), into billets.

The entire works on Walney Road now existed to supply its re-rolling mills.

By the end of the war, the works were both dated and inefficient. Despite the re-structuring of the 1920s, much of the infrastructure was still very much Victorian. With steelmaking becoming more and more competitive United Steel, in assessing the long-term prospects of the works, recognised that a process that eliminated the ingot/primary mill route

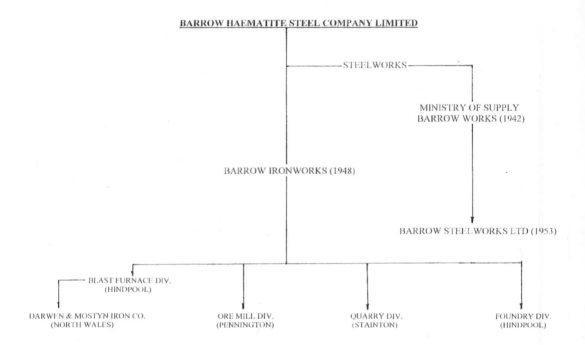

Splitting up the company in 1942 (S.Henderson)

would mean substantial savings [see *Key Stages of Steel Production* on p.3]. The principal attraction of continuous casting, apart from improved yield and reduced manning, was the enormous reduction in energy consumption.

Instrument technician, Ken Royall – later to be manager of the Fuel & Instrument Department - recalls that the first inkling anyone had of impending change was the posting, in 1951, on notice boards throughout the works, of an internal memorandum printed on Iron and Steel Board stationery, advising that Barrow had been designated an 'Experimental Works'.[1]

---

1 The memo was obviously alluding to continuous casting research.

Ray Millard, of Monk Street, Hindpool, an apprentice electrician at the time, recalls some of the preliminary electrical work put in hand prior to machinery installation. The old north-end casting pit had been earmarked for the site of the pilot plant and so new cables had to be run from the turbo house to the 2,400 KVA transformer.

"We had just finished replacing the old carbon-arc lighting in No.2 shed (carbon-arc had been the very earliest type of electric lighting in the 19th century, its replacement had been long overdue) when we received instructions to make a start on the power supply to the 5-ton arc furnace. Bill Jones, also from Hindpool, became involved, later becoming the Pilot Plant electrician. The route for the cables was through a very old part of the works and specifically through Bessemer Hill. This was a long gradient made of slag, rocks and ballast which carried a railway line along which ladles of molten iron were moved for charging the Bessemer Converters. It had lain disused since about 1919. While in the process of digging, the navvies made a discovery. Situated beneath Bessemer Hill, where it abuts the north wall of No.1 steel shed, they found a bothy, obviously used by the men of the Bessemer Shop. It contained tables and benches and items left behind by the men. The place had been sealed-up around 30-years earlier and forgotten about", explained Ray.

"Following the electrical work, the roof of No 2 steel foundry, together with the overhead crane track, was raised so that the vertical height from ground to the underside of the crane hook was over 46-feet. All of the related engineering work associated with the new pilot plant was undertaken by Barrow steelworks personnel. There were no outside contractors involved".

Members of the Electrical Dept. stood in front of the lower slope of Bessemer Hill. Bill Jones is on the front row, centre. Top left is the north-gable of No. 2 steel shed. The roof of No. 1 steel foundry (later the Boiler Shop), is on the right. Photo taken during cable- laying to the 5-ton arc furnace. *Courtesy of Ray Millard.*

Work on the new installation continued through 1951 – 52 then in December '52 the experimental work began. Also at this time attempts were made to gather information on progress made with the process at other works. No operational data could be obtained from the Low Moor Alloy Steel Works at Bradford, West Yorkshire who had started experimenting in 1946. It was as if theirs was being run as a covert operation! Intel gathered revealed that elsewhere, large sections were being cast, but at low speeds.

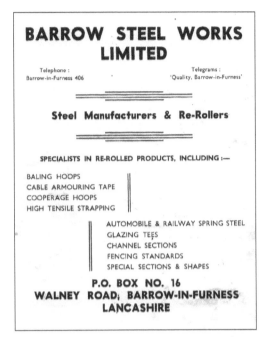

BARROW STEEL WORKS LIMITED

Telephone :
Barrow-in-Furness 406

Telegrams :
'Quality, Barrow-in-Furness'

Steel Manufacturers & Re-Rollers

SPECIALISTS IN RE-ROLLED PRODUCTS, INCLUDING :—

BALING HOOPS
CABLE ARMOURING TAPE
COOPERAGE HOOPS
HIGH TENSILE STRAPPING

AUTOMOBILE & RAILWAY SPRING STEEL
GLAZING TEES
CHANNEL SECTIONS
FENCING STANDARDS
SPECIAL SECTIONS & SHAPES

**P.O. BOX NO. 16
WALNEY ROAD, BARROW-IN-FURNESS
LANCASHIRE**

Barrow Hematite
and Semi-Phosphoric
Pig Irons available in
machine cast pigs to all
Steelworks and
Foundry specifications

"Barrow-Kinney"
Blast Furnace
and
Steelworks Plant

Special Machines
and
General Engineering
Work

Illustration (right)
shows 42" Plate Type
Goggle Valve Supplied by us

BARROW IRONWORKS LIMITED
BARROW-IN-FURNESS

Telephone : 830 (5 lines)   Telegrams: "IRONWORKS"

1950s ad for the works copied from the 6th edition of the Barrow-in-Furness official Handbook. The works had a frontage along Walney Road of 1 kilometre.

1950s ad for the now separate Ironworks.
Both images courtesy of Cumbria Archives and Local Studies Centre (Barrow).

Foremen's Council. Annual Dinner at the Wellington, Dalton-in-Furness on 7 December, 1957. Of those known, from the left are: Don Cottam; unknown; Ronny Metcalf; unknown; Mr T Marple.
Courtesy of K. E. Royall.

Presentation to works Ambulance Men. Of those recalled – 2nd from left is Mr McSweeney (Hoop Works); Ron Metcalf. 3rd from right is E. Ward, Hoop Works Mgr., 2nd from right – E. Aitkin, Works Mgr. (Note the Andrew's painting on the office wall – currently on display at Barrow's Dock Museum).
Courtesy of K. E. Royall.

Nationalisation Day (28th July, 1967) and a group of Barrow Works personnel are pictured with Lord Melchett, Chairman of the newly-formed British Steel Corporation. Among those known are, left to right:- Back row: unknown; Isaac Benn (Iron, Steel and Kindred Trades); Stan Beach Works Convenor (A. E. F.); Front row: Iain Halliday; Bernard Hogg (I.S.K.T.); Tom Maguire?; Don Cottam; Lord Melchett; unknown; C.D.A. Green, (Personnel Manager).
Photo courtesy of B. Hogg.

The northern-end of the works viewed from Walney Road across Cocken Lake. This photo was taken by Mail photographer, Fred Strike in 1963, at the time of the Iron Works closure. On the right of the image is the massive Sinter Plant building. *Courtesy of B. Hogg.*

A broad sweep of the northern-end of the works during demolition (c. 1984). Looking from left to right and across Cocken Lake we see the old ingot mould foundry; the electrical workshop with the water tower and turbo house behind. Next is No. 1 steel foundry (later the boiler shop) and then the Barcon with the goliath crane just visible. *Photo taken from the southern slopes of the slag bank by Brian Moxham.*

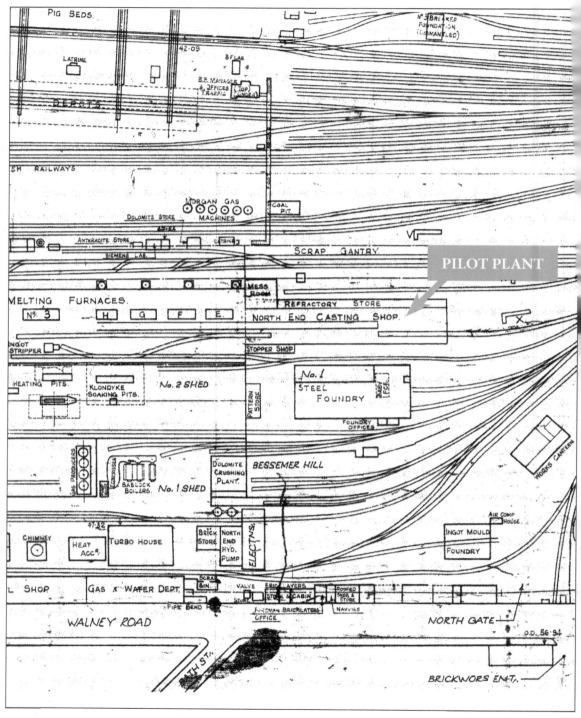

Part plan of the works showing the location of the Pilot Plant. Reproduced from Barrow Haematite Steel Company Plan No 72. *Courtesy of K. E. Royall.*

- LOCAL PERSONALITIES -

In his 1958[1] report to the Iron and Steel Institute, Iain Halliday acknowledged the efforts of the original Pilot Plant operatives, praising their tenacity and commitment, which undoubtedly contributed to the success of the project. However, for some reason, he chose not to mention his 'right-hand man', nor indeed others who also gave sterling service during the period. In his mind they were probably just doing their jobs.

William James Whiteside was born in Barrow and attended Barrow Grammar School. Upon leaving school in 1936 he started in the work's Analytical Laboratory, having been 'head hunted' by chief metallurgist, Joe Lyon. Apparently, Mr Lyon paid a visit to Head Teacher, Mr B. Johnson MA., and asked for his *top boy* in chemistry.[2]

Bill Whiteside (with pipe and pen parallel) at his desk in Barrow Steelworks. *Courtesy of K. E. Royall.*

Bill was involved in the new process right from the outset, transferring from the laboratory in 1952. His first task was to be sent on fact- finding missions to Sweden, where Barrow's new continuous casting machine was under construction.

---

1 Journal of the Iron and Steel Institute; 191 (1959)
2 Personal communication with Bill Whiteside Jnr

During May, 1952 he embarked for Copenhagen *en route* to the Swedish city of Malmo, the location of Broderna Edstrand AB. In the June of that year he went *via* Halmstad to the Nyby Bruk Steel Works where continuous casting had been, unsuccessfully, attempted during the early 1930s. His main recollection of Nyby was of how 'clean' it was! [Steelworks are notoriously hot, dirty, and smoky places].

Upon his return Bill reported his findings to Halliday. Later, following the successful discovery of Negative Strip in 1954, he was transferred to the fuel section of the Fuel and Instrument Department, which he managed until Barcon was commissioned [1961]. He became one of the shift managers there. During the 1950s Bill lived with his wife, Peggy, on Thorncliffe Road, Barrow. His hobbies were electronics; mechanics and photography, and also free-hand drawing. Upon leaving the works, they moved to Grange-over-Sands to run the Hampsfell Hotel. The couple worked well as a team, with Bill applying his steel industry management skills to good effect.

Whiteside gave commendable service and loyalty to Barrow Steelworks. He was indeed Iain Halliday's, unofficial, right-hand man, making a significant contribution to the development of Barrow's high-speed process. He left without ceremony, taking voluntary redundancy, in 1975.

Bill Whiteside (left), stood with Roy Millard who was one of the original Pilot Plant Operatives (Roy was the nephew of Albert Millard, manager of the Slag Reduction Company on Cocken Road). *Photo courtesy of Whiteside family.*

Francis (Frank) Pearson (7.9.16 – 5.11.73) was a Vickers Armstrong time- served engineer. He joined Barrow Steelworks in 1943, as a draughtsman, staying in the drawing office for two years before coming out as clerk of works for a reconstruction programme. He returned to the drawing office as chief draughtsman until August 1962, when he was appointed works engineer, responsible for all things mechanical. Frank was a key figure during the early 1950s being responsible for adapting the Rossi-Junghans machine[3] to Barrow Work's purposes.

The Engineer, Frank Pearson.
Courtesy of K. E. Royall.

Frank Pearson died suddenly in 1973, aged 57, following an operation at Barrow's North Lonsdale Hospital. He will be remembered for his problem-solving abilities and contribution to the development of Barrow's High Speed Process. Frank, whose hobby was gardening, lived with his wife, Peggy, in Longlands Avenue later moving to Kentmere Crescent, Barrow.

Continuous casting pioneer dies

ONE OF THE pioneers of continuous casting, Mr F. J. Pearson, works mechanical engineer at Barrow Works, has died suddenly at the age of 57.

Mr Pearson joined the steel industry in 1943 at Barrow Works to work in conjunction with their drawing office as clerk of works in the reconstruction of the steelplant at Barrow, which was undergoing considerable change due to the war effort.

At the end of hostilities he became chief draughtsman at the works, and was instrumental in the design of the continuous casting pilot plant. This was one of the first testing plants for continuous casting anywhere in the world.

Mr Pearson later concentrated his time and energies on the design and construction of the works continuous casting production plant as it is today.

It was at Vickers-Armstrongs, Barrow, that he served his apprenticeship in the company's engineering workshops and drawing office from 1931. He was with them up to joining Barrow Works.

He had been mechanical engineer since 1962. He was also a member of the metrication board, chairman of the safety and welfare committee, a member of the production committee, chairman of Barrow Works suggestions committee and a member of the works management committee.

The funeral service, held at Beacon Hill Methodist Church, Barrow, was attended by members of Barrow Works management and colleagues of Mr Pearson.

Steel News cutting re Frank Pearson. *Courtesy; Pearson Family*

3 In his 1958 report to the Iron & Steel Institute, Halliday stated that the reason why he omitted a description of the original machine, was because the re-configured machine bore almost no resemblance to the original.

Thomas Marple, previously works engineer, was to be the first manager of the newly formed Barrow Steelworks Limited in 1953.

Works Manager Mr. Marple (left), with Colonel Wingate. 1964. *Courtesy of K. E. Royall*

He was a local man who lived with his wife in a bungalow at Bardsea, near Ulverston. Mr Marple would preside over the change from orthodox steelmaking to the new continuous casting process. His hobbies included photography and he was a member of the work's camera club. Retiring in 1964 he was replaced by United Steels man, Mr E. Aitkin (the works had been officially acquired by United Steel combine the previous year).

Cllr Isaac Benn (Ike) one of the original pilot plant operatives. He was a senior shop steward and branch secretary of the Iron, Steel and Kindred Trades Association. Ike first joined the company in 1938 but after war service became a member of the Birmingham City Police. Coming back to the company in 1949, he worked on the Cold Reeler[4] in the Bright Bar Department. In 1951 Ike volunteered to work on the Pilot Plant. Apart from a short stint as work's security guard in the early 1980s he stayed with Continuous Casting for most of his career.

---

4 Cold Reeling was just one way of finishing rolled product. As an example, round bar destined for making shell casings during the war was subject to centreless grinding. This gave a highly polished finish (hence bright bar). Subsequent cold reeling would then remove any marks left behind from grinding. Incidentaly the centreless grinding machine made an unholy noise – akin to a wailing banshee – much to the annoyance of the Scotch Building residents of Walney Road.

He became mayor of the Barrow Borough, 1981-82. He lived with his wife and daughter at 6, Tyne Road, Walney Island, Barrow.

Joseph Lyon MBE[5] spent all his working life at Barrow Steelworks. He was chief metallurgist during the United Steel years. Born in Barrow he attended Barrow Grammar school and then gained his degree at Manchester University before going into the works as a junior chemist. Joe's elder brother, James (*Smut*) Lyon, established Lyon's chemists in Hartington Street, Barrow in 1933.

Councillor Isaac Benn in his trappings of office. Photo: Barrow Town Hall, *Courtesy of Susan Benson.*

Mr J. Lyon, Chief Metallurgist. *K. E. Royall.*

Mr Lyon's office was located in the work's Test House. He had full responsibility for independent inspection (Quality Control). Sadly, while in his sixties, he collapsed and died at work. He was succeeded by Joseph Harry Brown, who had come to Barrow as a junior chemist, from Steel Peech and Tozer (Rotherham), in 1943.

Joe Lyon was awarded the MBE for his part in the development of the high speed process.

---

5 Mr Lyon's MBE could not be verified via the *London Gazette*. His nephew, Derek Lyon and the town's one-time Chief Executive, (Town Clerk), assured the authors that he was in receipt of this honour.

The six original pilot plant operatives snapped in 1951.

Left to right: Isaac Benn, Bill Marklew, David Haney, Bill Caldwell (Bill had been a key figure and went on to become a Barcon shift manager); Bert Calderbank? , Roy Millard, unknown. *Courtesy of Caldwell family.*

- CONTINUOUS CASTING -
BARROW-IN-FURNESS, LANCASHIRE, UK
(STRANGGIESSEN)

1. DEVELOPMENT

Preamble – British Developments

After the Second World War, development in Britain was confined mainly to laboratory-type equipment in certain universities and at the laboratories of the British Iron and Steel Research Association, BISRA. In 1946, however, the Low Moor Alloy Steel works Ltd in Bradford, West Yorkshire, installed a 2.5-tons pilot plant which has since been operated on a production basis. In 1952, experiments began at Barrow with a 5-ton capacity plant of the Rossi-Junghans type, from which the high-speed Barrow process has been developed.

In 1954, a syndicate of ten Sheffield steelmaking companies installed a 30-cwt plant at the works of William Jessup and Sons Ltd, in Sheffield. This plant was of a type developed by BISRA and, although it ceased operations in 1957, BISRA had, in 1955, installed an experimental machine at their Sheffield laboratory.

Developments in Barrow progressed to the stage where, in 1957, the United Steel Companies increased the capacity of the plant to 7.5-tons and, in 1958, rebuilt the machine as a twin-strand unit with automatic operation. The progress and level of development of the process at Barrow may be judged from the fact that, of the 29 installations (1959), in operation globally, 12 were based on the Barrow plant.

(Paraphrased from Steel Review, July 1959)

Introduction

The continuous casting process for steel developed at Barrow during the early 1950s is characterised by high casting speeds. The small section billets of mainly 2-inches (50mm) square were cast at tonnage rates of 7 – 8 tons per hour. The process was initially developed on a single mould machine supplied with 2 and 4-ton amounts of liquid metal. Later, the ladle capacity was increased to 8-tons when the machine was converted to a twin-strand unit casting at 15-tons per hour.

All steel fully killed[1]; yield: 98.5%; product quality: good.

Background

Despite a glorious *Bessemer* past the Barrow Haematite Steel Company, by 1940, had insufficient resources to bring the works into the 20th century. During 1942-43 the works came under the control of the United Steel Companies of Sheffield who set in place some long overdue improvements to the open hearth plant and supporting services. In assessing the long-term future of the works, United Steel recognised the advantages to be gained if a process could be developed to produce, direct from liquid metal, the small section billets and slabs normally supplied to the five re-rolling mills. (50 up to 100mm square billets and 150 x 50mm slabs). Qualities included the carbon steel range up to 0.08% C. and also certain low-alloy specifications, e.g. – silico-manganese and manganese molybdenum spring steels.

It was decided to explore the earlier work of German metallurgist, Siegfried Junghans, who had used an open-ended mould to cast copper and brass. Running into difficulties with the solidifying metal sticking to the sides of the mould he arranged to oscillate or reciprocate the mould, which was done at Barrow but with a significantly different mechanical motion. In 1933 Junghans patented his process. Working in Europe as an investment banker, entrepreneur Irving Rossi met with Junghans and immediately saw the potential if the process could be used for steel. He acquired exclusive rights to the Junghans patent in the USA and England. In return Rossi (who would later form *Concast AG*) agreed to finance the commercialisation of the process outside Germany.

Available Intel (1951)

Continuous casting had been used commercially for some time in the non-ferrous field e.g. for copper and brass and aluminium. Developments for steel however, were still at the pilot plant stage and, as far as was known, centred around the following concerns: (see table on p23)

Nothing was known, in 1951, of operations at Jacob Holtzer, Unieux, France. Additionally, no data could be obtained regarding the Low Moor Alloy Steelworks in Bradford, Yorkshire, UK.

Available information showed that experience thus far was concerned with the casting of large sections at low casting speeds, in the range of 24 – 36 inches per minute. **Nowhere had the sizes required at Barrow**

1 By 'killed' is meant oxygen killed i.e all oxygen removed

Country	Started	Company	Plant	Plant Capacity
Germany	1943	Siegfried Junghans	Schorndorf	1.5-tons
	1950	Mannesmann-Hüttenwerk[5]	Duisburg-Huckingen	5-tons
Austria	1947	Gebr. Böhler AG	Kapfenberg	0.3-tons
USA	1946	The Babcock & Wilcox Company	Beaver Falls, Pa.	5-tons
	1949	Allegheny-Ludlum Steel Corp.	Watervliet, N.Y.	3-tons
UK	1946	Low Moor Iron & Steel	Bradford	2.5-tons
	1952	United Steel Companies	Barrow	5-tons

been cast on a routine basis. (For any method to be economically worthwhile, it would be necessary to achieve casting speeds up to ten times faster than any speeds then known).

Iain M. D. Halliday, a United Steels Ferro-metallurgist from Sheffield was appointed to Head of Continuous Casting Research. He had previously been instrumental in upgrading Barrow's Open Hearth Plant. Industrial chemist, Bill Whiteside of Barrow works was transferred to the project while Frank J. Pearson, who joined the company in 1940 from Vickers Armstrong, Barrow, led the engineering team.

Terms of Reference

In September, 1951 a steering committee was established which consisted of United Steel and Barrow Steelworks personnel. At the first meeting, where terms of reference were agreed upon, and targets set, Iain Halliday stated that United Steel had confidence that Barrow could proceed from a sound basis of know-how and intelligent expe-

rience gained over 70 years[2]. He added that the process which they hoped to develop had been used for some time in the non-ferrous field, but progress with steel was negligible. (Copper, with its higher coefficient of thermal conductivity, lent itself to the process whilst steel, with a much higher melting-point, had a much lower thermal coefficient). Process development *via* works trials was envisaged. Many problems were involved – for example, techniques for liquid metal handling; for control of casting and product quality and even the simple question of how to pour liquid steel cleanly into a 50mm square aperture mould. During 1951 contracts were agreed with Rossi for the rights, as licensees, to use the *Rossi-Junghans* process.

The Plant

The pilot plant was installed towards the northern-end of the works in what was once the No 2 steel foundry and comprised of a 5-ton (later increased to 8-ton) basic arc furnace. This location meant that

---

2 The fact that the Barrow works had an extensive engineering support facility would also be a deciding factor. Continuous Casting would be as much a challenge for engineers as it was for metallurgists.

site conditions were suitable for using open-hearth metal from one of the 40-ton furnaces that were still in operation. An electric furnace was the obvious choice as steel for continuous casting requires tighter control than for orthodox ingot practice, where bad furnace operation can be corrected in the ingot mould. Also, slag could be removed from the arc furnace before tapping; (Fig. 5). In fact a *double-slag* procedure was adopted. With the furnace type agreed, ladle designs were the next consideration. Up to this time only bottom-pour (stoppered ladles) had been used at Barrow. These presented difficulties with flow control (due to the reducing *hydraulic head* as the ladle emptied) and did not lend themselves to pre-heating as the stopper-rod became overheated. The risk of a 'running

stopper' could not be taken. Frank Pearson's technical people came up with the 'semi-teapot ladle'. This design could be preheated and allowed pouring to continue for nearly 2 hours. Ladle pre-heating was achieved via a combustion-can, oil-fired burner whilst the ladle sat on top of the tower, (in a production plant setting it was envisaged that purpose-built pre-heating stations would be used). The tower (Fig.1) and contained machinery were juxtaposed the arc furnace with a 2,400 KVA transformer.

Local management co-operated with Mr Rossi in the design and construction of the original machine but disagreeing with him on the positioning of the furnace. Rossi intended for the furnace to be sited on top of the machine

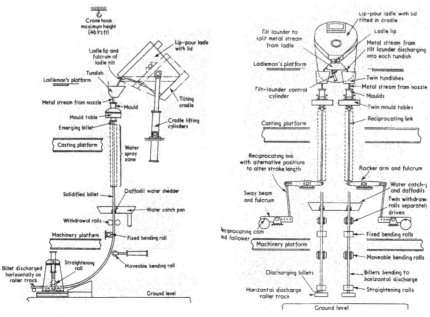

Vertical disposition of casting machine elements. Side-view (left); then front view.

tower, Barrow insisted the furnace be at ground level. As built, the machine had certain facilities for twin-stranding. The initial tundish installation comprised a pair of tilting tundishes set side by side and designed to nozzle- pour into two moulds mounted on a common support table and reciprocated by means of a single cam driven through gearing by a 5-hp motor. This motor also drove a single horizontally placed withdrawal roll common to both strands. A small undriven withdrawal roll was arranged, when spring-loaded, to press against the driven roll for each strand of the machine. The initial discharge equipment comprised a horizontal traversing torch and pneumatically-powered tilting frame carrying a movable basket to eject the cut billets horizontally. (The machine was later modified (21st March, 1956) to bend the billets along a roller-path).

The Process

Prior to the first trial a thorough risk assessment was conducted. Job hazard analysis was standard policy when tackling new operations and proved extremely valuable in the establishment of safety precautions. Previously, at Barrow, all operations involving molten metal had been carried out over a pit; even furnaces were constructed over a pit so that any *melt-down* would see the liquid steel dumped out of harm's way. Now, men were being asked to operate almost underneath a cauldron of molten metal which would be teeming said metal towards them. In addressing the risk operatives were attired in asbestos suits, heavy leather boots and helmets fitted with visors and mesh. Additionally, a reservoir of water was sited overhead fitted with a chain mechanism which could be pulled, like a toilet cistern, thus releasing the water over the operatives. (This facility was never used – thankfully) The first trial was undertaken on Tuesday, 2nd December 1952 with a casting size of 3-inches (75mm) square. On 23rd February 1953 the first attempt was made at casting a 2-inch (50mm) billet – like the first, this was unsuccessful. These early trials were done with a metal charge of only 2-tons and although not entirely satisfactory, advances were made and experience gained in other areas such as pre-heating practice and hot-metal handling. Commencing to cast with the steel temperature too high resulted in *breakouts*; with the temperature too low the metal would *freeze* in the receptacles. The optimum temperatures were arrived at empirically and then a series of encouraging trials followed. The problem of how to pour liquid steel cleanly into a small aperture mould was solved at Barrow. This was achieved by the design of a *non-swirl* nozzle made of *Zirconia*. These nozzles were set into the magnesite brick of the tundish and comprised a *castellated* top which effectively disrupted the vortex. The metal poured out like a solid rod (12mm) diameter into the mould.

During 1953 the first attempt was made at automatic control. In January 1953, following many experiments with mould reciprocation, the *alchemy* was lighted upon. What was termed *negative strip,* then later the *Barrow Principle,* became established enabling faster casting speeds. In summary,

the mould was reciprocated vertically, as done by Junghans, but with the important difference that *negative strip* was introduced. This was achieved by moving the mould downwards slightly faster than the descending billet for three-quarters of the reciprocation cycle. In each remaining quarter-cycle, the mould was raised relatively quickly – at about three times the descent speed. *Negative strip* reduced drag and prevented sticking between the newly cast billet skin and the mould faces – thereby allowing high-speed casting. It also created conditions whereby surface cracks, caused by temporary lack of lubrication, were welded-up.

Date	Trial No	Size	Detail
5th May, 1953	54	2-inch (i. e. about 50 mm x50 mm)	Metal charge increased from 2 to 4-tons
8th January, 1954	181	2-inch	Negative Strip established
12th January, 1954	183	2-inch	First trial with open-hearth metal
11th June, 1954	267	2-inch	First attempts at high-speed casting
7th July, 1954	281	2-inch	Max. speed attained: 570-inches per minute
21st September, 1954	420	6 x 2-inch	First 18/8 stainless steel slab cast (quality excellent)
29th August, 1956	547	3-inch (i. e. about 75 mmx 75 mm)	Use of aluminium feeder to de-oxidise steel
18th April, 1957	694	3-inch	First trial with 7.5-ton ladle charge
26th September, 1957	753	3-inch	Gamma-ray equipment tests for auto-control
22nd November, 1957	785	2-inch	Complete cast run under automatic control

Tab. 2:   Significant Milestones, single-strand machine

Date	Trial No	Size	Detail
6th May, 1958	788	2-inch	First twin-strand machine trial
23rd June, 1958	803	2-inch	First complete twin-strand cast under automatic control
12th September, 1958	863	3-inch	Twin-stranding of 3-inch billet begun
19th September, 1958	869	3-inch	Last cast before reconfiguring to slab machine

Tab. 3:   Significant Milestones, twin-strand machine

Once the elusive *negative strip* had been established speeds in excess of 200-inches per minute could be achieved. Then for some reason *breakouts* began occurring below the mould at these higher speeds. Investigations traced the problem to the construction of the mould; details of which had been supplied by Rossi. The Junghans mould was of solid copper, 32-inches long, assembled in two halves giving a longitudinal seam. Although almost a perfect fit, this seam, under certain conditions, gave rise to very fine fins being formed on the two opposing faces of the billet. These fins were causing the problem. The mould was redesigned to be one piece of solid copper, 0.25-inch (6mm) thick with the addition of a chrome-plated internal finish. Trials with the Barrow design of mould allowed greater and sustainable speeds and for the first time it was possible to completely drain the ladle through the machine. On 31st March 1954 over 1000-feet of 2-inch billet was produced in one run. Then on 7th July a speed of 405-inches per minute was achieved. ***This was to be the fastest <u>average</u> speed for casting a 2-inch billet.***

During 1954 industrial chemist, Eric Grayson was appointed assistant plant manager.

Mould deck of the twin-strand machine. Top right is the control pendant while cooling water pipes to the moulds can be seen left and right.
Courtesy of N. W. Evening Mail.

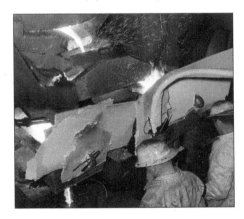

Twin-stranding on manual operation. This view shows the metal stream leaving the ladle lip into the launder where it splits right and left into the tundishes. At bottom left the metal stream can be seen entering the 2-inch square aperture of the reciprocating mould watched by the mould operator, who is controlling the casting speed. *Courtesy K.Royall*

Taking the bath temperature of the 5-ton arc furnace

During 1958 Mr A. Jackson, Technical Advisor on Steelmaking to the United Steel Companies, was co-opted onto the steering committee. Now that production runs could be made routinely with 2-inch billets, other sizes were trialled – all proved successful. Towards the end of 1958, after a total of 869 trials over 6-years, it was concluded the development work was successful; finance could now be released for the laying down of a production plant at Barrow. In the meantime the pilot plant was adapted for trials with large 36 x 5.5-inch slabs in a variety of steels including chrome-nickel stainless steel (cast at 50-ins per minute).

During the early 1960s Halliday had success with the continuous casting of various high-alloy steels. This necessitated the shrouding of the metal stream with lambent flames from propane gas. He submitted patent applications for this work.

Barrow works now became a training ground for other members of the United Steel Group of companies; starting with five men from the Appleby Frodingham works at Scunthorpe (Fig.7). This massive plant at the time was arguably the largest integrated steelworks in Europe. After running for some time on a semi-production basis, the Barrow pilot plant was de-commissioned (1969).

The first group from Appleby Frodingham, Scunthorpe, wearing their protective suits and looking pleased with their progress, at the Barrow plant in 1960.

2. PRODUCTION

Following an official announcement on the 8th June, 1959 work started on the building (Fig.10) that would house the new production plant at Barrow works. The new plant – *Barcon* – would comprise a 25-ton electric arc furnace and two twin-strand casting machines of the vertical with bending type. The new building would lie on a north-south axis and align the old Siemens open hearth melting shops and No 3 rail mill. It was anticipated the unit would provide 36 to 46,000-tons of billets per year thus making Barrow the first works to rely on continuous casting for its entire production. (In subsequent years the output reached 50,000-tons). The plant ran on a 21-shift week.

When plans were first issued for the project they were annotated *stage 1*. It was intended that a second stage would see an extension, with the hoop and bar mills, which were about one mile distant, resited adjacent to *Barcon;* creating what became known later in the USA as a *mini-mill.* This for some reason never materialised (possibly because of the impending nationalising of the industry). *Barcon* stage 1 cost £1.35 million pounds. No record can be found regarding the cost of the six-year development work.

External view of the *Barcon* Production Hall viewed from the south of the works. This was erected by civil engineering contractor *Wimpey* during 1960-61.

The mould used for large slab casting; note the thermocouples for temperature monitoring, also the rapeseed oil pipework.

Continuous cast chrome nickel stainless steel slab, 36-inches x 5.5-inches. At the time this was the largest size produced by continuous casting in the UK. Circa 1959.

As reported in *New Scientist* of 12 May 1960 many countries, including Belgium, France and Russia were keeping a careful eye on developments in Barrow. The report of the 1958 meeting of the Iron and Steel Institute regarding the successful development work and the establishment of negative strip – soon to become normal industry operating practice – had aroused a great deal of interest within the industry worldwide. Barrow had made history in more ways than one; as well as pioneering the high-speed process it was also the first steel works to operate without primary mills. Another major development in steelmaking during the 1950s and 60s was the use of oxygen. Barrow works exploited this introducing

be exercised as overlong use raised the metal temperature too high, resulting in excessive wear of the refractory lining. The average life of the lining was 193 heats against 281 before lancing.

While the furnace was producing a heat, the components involved in the process were being preheated by special burners to a predetermined temperature. At the furnace, for a given metal analysis, a metal tapping temperature was obtained and checked by immersion thermocouple. When ready the furnace was tapped into the ladle, which was suspended into the tapping pit by the 50-ton crane. From here the ladle was transferred to the scales for weighing and then lifted

Oxygen lance and oil burner attached to the boiler-plated side of the arc furnace; it had previously been portable and deployed via the furnace fire-door.

A ladle-man taking the metal temperature prior to casting. 19th Oct 1961

oxygen lancing to the electric steel-making of *Barcon*, firstly deploying the lance via the furnace fire-door and then as a permanent fixture mounting it on the boiler plated side of the vessel (Fig.11). Care had to

onto the top deck of the machine. The metal temperature was again checked (Fig.12); if too high a cooling period was allowed; if optimum the combustion-can oil burner was attached and ignited. At this point the mould operators, who were in contact with the ladle man *via* intercom, turned on the mould cooling water and rapeseed oil pump, then hung an index wire 14-inches long into the moulds. Below the mould deck operatives were busy feeding the dummy bars up through the withdrawal rolls to the bottom of the open-ended moulds. Dummy bars were the same cross-section as the section about to be cast. The top of these dummy bars were fitted with protruding bolts secured by tapered pins, which, together with asbestos washers, closed off the bottom of the moulds

To initiate casting manually the mould operators instructed the ladle man to tilt the ladle forwards, the steel established a steady stream into the transverse launder then *via* a tundish into one of the moulds (the tundish was so designed to allow one strand to be started initially, once casting was established the other strand would be started). The mould operator watched as the metal rose up inside the mould, when it reached the end of the wire the machine would be started by pressing the start-button on the control pendant. The moulds began to reciprocate and the withdrawal rolls pulled the dummy bars down followed by the newly cast billets into the water-spray cooling chamber. Operatives at ground level were warned, *via* the intercom, to discon-

nect the dummy bars by knocking out the tapered pins. They next activated the hydraulic bending rolls to direct the new billets through the straightening rolls and along the discharge roller path where they were cut to length by flame-cutting equipment.

While casting was ongoing the furnace would be re- charged for the next heat and the process started all over again. At this point if we pause to consider the word *continuous*, which means without end or unceasing, we see that as far as the process at Barrow was concerned the word was a misnomer. Casting took over two hours but the furnace took four hours to produce a melt. Had stage 2 of *Barcon* been implemented then the process would have been truly continuous,

Following the nationalisation of the industry in 1967 Barrow works became part of the General Steels Division – one of the six new product divisions of the BSC. Barrow was linked with Workington as the *Teesside and Workington* group of the new division. Continuous casting did little for the Barrow economy. It did, however, secure employment for about 1000 men and women for a period of 20-years. On 23rd, November, 1983 the works, after years of uncertainty, finally closed. Nothing exists on the original site to remind us of a once proud industry.

25-ton arc furnace under construction. The pilot plant furnace was similar, but smaller. *Courtesy of The Mail*

Pre-heating a ladle in one of the purpose built preheating stations.

'Another Good Tap'. Charged ladle about to be lifted out of the tapping pit viewed from the overhead crane; the two casting machines can be seen at top of photo, 26th July, 1962.

Charged ladle being transferred to the upper deck of one of the casting machines.

With the ladle secured into the cradle of the west machine and burner attached, casting is commenced.

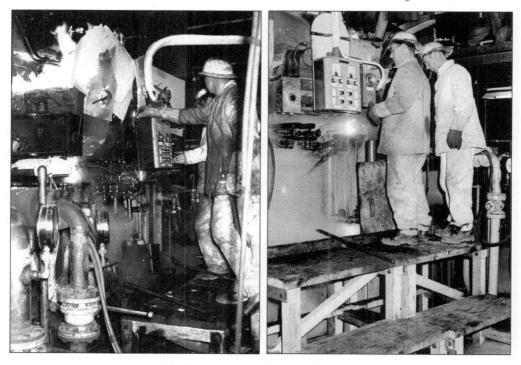

Mould operators controlling the casting speed.

Close-up view of the process; metal can be seen emerging from the ladle (left) into the transverse launder and then into one of the tundishes. 26th July, 1962.

Left - newly cast billets passing through water spray chamber.
Right - billets pass through withdrawal rolls.

Twin strands coming off the bottom of the machine turning through straightening rolls and up the discharge roller path (left).

The straightening rolls (below).

The cooling and inspection bank with newly cast billets on skids (looking north)

Roller discharge path and billet platform/tilting table with hot bank beyond. A total of 76 in number 1.5 HP Radicon electric motors (38 each side) propelled the billets up the discharge path to the hot bank.

Reciprocating cam and follower, sway beam and fulcrum (mould reciprocation mechanism)

Withdrawal rolls and motors

View of casting machines from the billet discharge bay, following installation of personnel lift (centre); the west machine is casting. 1961

SOURCES

Steel Review (1959)

Continuous Casting at Barrow (1958); I.M.D. Halliday B.Sc

Continuous Casting of small Billets at Barrow Steel Works (1964);

A. Jackson, J.Lyon, T. McGuire

The Continuous Story of Continuous Casting (1975),

British Steel Corporation.

Personal Communications with plant operatives.

All photos, unless stated otherwise, are courtesy of K. E. Royall.

- The United Steel Companies -
of Sheffield

United Steel was formed in 1918 by the merging of shares of the UK's largest steel plants, later acquiring the Appelby Frodingham works at Scunthorpe also the Workington Iron & Steel Company. United Steel were at the forefront of investment and innovation. Through heavy investment they had improved the efficiency of iron making by the use of sintering. In their Seraphim blast furnace the entire burden was sintered. Following the end of the war the company spent over £100 million on capital development. Barrow and Workington were on the receiving-end of this investment. Throughout the 1950s the combine employed over 38,000 people.

During this decade United Steel invested heavily in improvements to the Open Hearth Process at its Swinden Laboratories in Rotherham. Built after the Second World War, 240 specialised staff were employed in research and development.

It was also during the fifties that their attention began to be focused on Continuous Casting.

Subsidiary Companies, UK

Among its UK subsidiaries were the Beckermet Mining Company, Workington, Cumberland; Distington Engineering Company, also at Workington and the Yorkshire Engine Company Ltd., Sheffield.

Overseas Subsidiaries

These included companies and organisations in Canada; India; Iran; Pakistan; South Africa and Switzerland.

Albert Jackson A. MET. F.I.M., CGIA

Albert Jackson started work as a shift chemist in the open hearth shop at the Frodingham works, his energies enabled him to gain a measure of scientific learning, and early in his career he attended Sheffield University where he gained an Associateship in Metallurgy. He was a Fellow of the Institute of Metallurgists. For many years he was the Chairman of the Iron and Steel Operatives Advisory and Moderating Committees of the City & Guilds of London Institute.

Albert Jackson, he was the main player with regard to United Steel's Ajax Furnace project.
Courtesy of Ken Royall.

Early in 1958 it was announced that Mr. Jackson had been appointed Technical Advisor on Steelmaking to the United Steel Companies. It was also in 1958 that he was co-opted onto the Continuous Casting Steering Committee at Barrow-in-Furness.

- LOW MOOR ALLOY STEEL WORKS -
(REF: LOW MOOR IRON WORKS, BRADFORD. CHARLES DODSWORTH)

Sited about 3 miles to the south of the Yorkshire city of Bradford, the works were originally known as the Low Moor Iron Works. Built in 1791 the works, actually a wrought iron foundry, were erected to exploit the high-quality iron ore and low sulphur coal found in the area. At one time it was the largest iron works in Yorkshire and had various owners over the decades.

The last recorded owner (from 1928), was Thos. W. Ward[1] who modernized some of the plant.

At some point during 1946 a small plant started experimenting with continuous casting. Unfortunately Iain Halliday (in 1951) was not successful in establishing exactly what their terms of reference were, nor what progress had been made.

> *When first the shapeless sable ore,*
> *Is laid in heaps around Low Moor,*
> *The roaring blast, the quiv'ring flame,*
> *Give to the mass another name:*
> *White as the Sun the metal runs,*
> *For horse-shoe nails, or thund'ring guns. . .*

> Airedale Poet, John Nicholson.

---

1 The same Thos. Ward who ran the ship-breaking operation at Barrow's Devonshire Dock

- GALLERY -

Barrow Pilot Plant: Single-Stranding in 1954. The newly-cast billet can be seen emerging from the bottom of the machine tower (Centre) and progressing along the discharge roller path where it will be cut to length. *Courtesy of K. E. Royall.*

Barrow Pilot Plant: Rear view of the 5 ton furnace being charged with scrap steel.

A furnace hand is fettling the firebrick lining of the furnace launder

The furnace has been emptied into the preheated ladle which is suspended from an overhead crane.
Images courtesy of K. E. Royall.

The Tap: The arc furnace tilts and molten metal pours into the waiting, preheated, ladle which is suspended by an overhead crane in the tapping pit. 1954. *Courtesy of K. E. Royall..*

The area underneath the casting tower where an operative is using an oxy-flame torch to cut the billets to length. *Courtesy of K. E. Royall. 1956.*

Single-Stranding in 1956. The newly cast billet is emerging from the cooling chamber then bending into the hydraulic operated straightening rolls prior to being cut to length

Further along the roller path an operative cuts the billet to length watched by Dabber Haney (Left).
Both images courtesy of K. E. Royall

Back Row: LtoR - Bill Marklew, Unknown, Jim Chelton, Jack Skae, Gordon Instance.
Front Row: Danny Smith, Unknown, Jack Lowther. *Courtesy of Instance family.*

Group photo of Pilot Plant operatives and associated disciplines following a game of football. This image is significant to the author in that it features friend and ex-colleague – Draughtsman Jack Smithson (front row, left). *Courtesy of Instance family.*

- MISCELLANEOUS -

Bill Whiteside, left, with his Swedish tour guides waiting to dine in Copenhagen. May, 1952.

The Grand Hotel at Halmstad where Bill Whiteside stayed, June, 1952. Both images courtesy of Bill Whiteside jnr

TELEGRAMS: VICASTRONG, BARROW-IN-FURNESS.

TELEPHONE: 351 BARROW (8 LINES).

Vickers-Armstrongs Limited.

Naval Construction Works,
Barrow-in-Furness.

Your Ref.

Our Ref. WM/TO.

Enc

22nd May, 1943.

<u>TO WHOM IT MAY CONCERN.</u>

I hereby certify that Mr. F. J. Pearson entered these Works as an Engineering Apprentice in 1935 and obtained three years practical experience in the Boiler Department before entering the Marine Engine Drawing Office in 1936 as the result of a competitive examination. Here he was engaged, particularly in the Boiler Section, on arrangement and detail drawings of machinery installations for Naval and Merchant vessels constructed by this Firm. As a member of our Record Party he has had opportunities of observing the machinery in operation at sea.

He has also had considerable experience in connection with the extensions and repairs to the buildings and plant of these Works, some of which have been major undertakings, to meet the heavy commitments upon our resources due to the War.

Apart from the work upon which he has been directly engaged he has had opportunities of gaining valuable engineering experience in connection with the wide range of products manufactured by this Firm.

He has satisfactorily carried out all the work which has been entrusted to him as a member of my Staff and, in leaving us to undertake work of national importance elsewhere, he has my best wishes for his future career.

CHIEF DRAUGHTSMAN,
MARINE ENGINE DEPARTMENT.

Vickers – Armstrongs letter of 22 May, 1943 relating to Mr F. J. Pearson. *Courtesy of Frank Pearson jnr.*

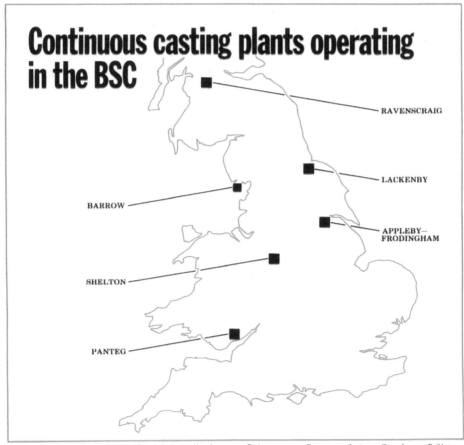

Continuous casting plants operating in the BSC

RAVENSCRAIG

LACKENBY

APPLEBY— FRODINGHAM

BARROW

SHELTON

PANTEG

	Commissioning Date	No. of Strands	Design	Furnaces	Section	Type of Steel	Builder
GENERAL STEELS DIVISION							
Teesside & Workington Group BARROW	1961	2 × 2	Vertical with bending	Electric arc	Billets	C and low alloy	Distington Eng.
South Teesside Works LACKENBY	1972 1973	8 2	Concast S Concast S	B.O.F. B.O.F.	Blooms Slabs	C Steel C Steel	Distington Eng.
Scunthorpe & Lancashire Group APPLEBY—FRODINGHAM WORKS	1973	2 × 2	DEMAG/Mannesmann Curved Mould Constant Radius	B.O.F.	Slabs	C Steel	Principal UK sub-contractor & builder, Davy & United Eng
Scottish Shelton & East Moors Group SHELTON WORKS	1964	2 × 3 3 2	A. Vertical with bending B. Vertical C. Vertical	Kaldo	Blooms Blooms Slabs	C and low alloy	Distington Eng.
STRIP MILLS DIVISION							
Scottish Group RAVENSCRAIG WORKS	1974 1975	1 1	Concast S Concast S	B.O.F. B.O.F.	Slabs Slabs	C Steel C Steel	Distington Eng.
SPECIAL STEELS DIVISION							
Alloy & Stainless Works Group PANTEG	1962	1	Vertical	Electric Arc A.O.D.	Slabs	Stainless	Distington Eng.

Distington Engineering Company. U.K. licensee for the building of Concast machines. Other products from BSC's Distington Works include:—
Iron Foundry. Ingot moulds, bottom plates, slag pots etc.
Non-Ferrous Foundry. Copper castings for blast furnaces, oxygen lance nozzles, arc-furnace electrode clamps.
Engineering. Charging ladles, ladle cars, torpedo ladle cars, slag ladle carriages. Roll stands, dressing and straightening equipment. Fuel element flasks.

B.S.C. (International) Limited hold a 22% share of the equity of Concast A.G., Zurich. Cmdr. M. G. Lyne, General Manager, Distington Works of the Special Steels Division is a member of the Board of Concast A.G.

Map showing the works operating CC plants in 1975. Copied from a British Steel promotional pamphlet.
How many of them are still in existence?

- THE BARROW PRINCIPLE -

The Barrow Principle, originally known as 'Negative Strip', was the once elusive mould motion that allowed for casting speeds of more than 200 inches per minute (For a certain period 'Positive Strip' was tried – obviously with no success). The mould was moved up and down via a cam and link mechanism driven by an electric motor. The reciprocation cycle was such that on the down stroke the mould travelled slightly faster than the section being cast. It then returned on the upstroke about three times as fast as the descent speed. This slightly faster descent speed was the main characteristic of the Barrow Principle.

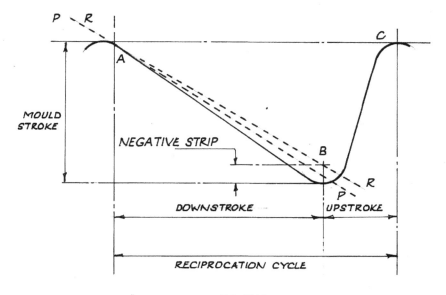

Diagram of Mould Motion

For all casting speeds the movement of the billet can be superimposed on a diagram of mould movement, which basically is the profile of the (3 in 1) cam. This can be seen on the diagram. The line ABC is the mould movement to a time baseline of one complete reciprocation, as represented on the cam profile. Between A and B the mould descends through its own stroke length at nearly constant rate for three-quarters of the cycle, whilst between B and C the upstroke occurs. The dotted line PP represents the superimposed descent of the billet, if conditions are synchronised, as is done when operating on the Junghans Principle. The line RR represents the billet descent when slower than the mould. Since these line coincide at the beginning of mould descent, the difference at the end is real and, as indicated, is known as negative strip.

In 1958 this mould motion was captured on a cine film which is now with the BFI.

- Barcon Production Plant -
Manning pattern during the early 1960s

	Plant Manager: Tom Maguire			
	Yellow Shift	**Green Shift**	**Blue Shift**	**Red Shift**
Shift Manager	Bill Whiteside	Jack Morecambe*	Bert Jackson	Bill Caldwell
Ladle	Ernie Guy	Bill Henderson	Billy Fell	Jim Johnson
Mould	Dennis Chelton	Ike Benn	Jack Johnson	Bill Chelton
	Tom Grisdale	Geo. Warriner	Stan Henderson	Geo. Nelson
Cut Off	Ron Walmsley	Gordon Finlayson	J. Johnson	D. Haney
	Walter Rutnowski	Jan Dubka	Geo. Bennett	Jack McMahon
Panel	B. Hogg	Nobby Clark	Norman Bircher	Eric Whitehead
Arc Furnace				
Leading Hand	Bill Mulholland	Jack Gardner	Fred Shepherd	Tom Lightfoot
	Tom Evans	Jim Caldwell	Bob Carter	Ronnie Whitehead
	3rd Hand ?	3rd Hand ?	Billy White?	Wilf Jones
Cranes				
50-ton	Bert Greaves	Colin MacDonald	Billy Hunter	Harold Johnson
25-ton	Frank Green	Jan Syzmala	Billy Lamb	Ernie Asbury
Goliath	Bert Greaves			
Utility Men	John Green	Mike Williams	Jimmy Deakin	Mike Brown
Chemists	Rodger Bradley	Walter Scott	V.Hall	D.Anderson
	Frank Wignall	John Miller	Doings	Doings
Dipper Men	A. Finch	Bert Logan	Dick Walkden	J. Irwin
Toilet Attendant	Jimmy Carson			

* Later superseded by Dennis Chelton

The four shifts worked a 48-hour week over 6 days with 2 days off. Towards the end of the 1960's a Continental Shift system was adopted whereby a man worked 2 nights (10pm to 6am); 2 afternoons (2pm to 10pm) and then 2 mornings (6am to 2pm). Followed by 2 days off. To be elegible to work on Barcon you had to be 21 years-old i.e. an adult.

Blue shift c. 1962 in the mess room. From left to right are: Bert Jackson (standing); Fred Shepherd; Billy Fell; George Bennett; Harry Lamb; Unknown; Jim Deakin; Billy Hunter; Stan Henderson; Jack Johnson; Unknown visitor. *Courtesy of Billy Fell.*

View from Barcon's 25-ton crane showing the arc furnace, fully tilted, after being tapped into the ladle. The ladle is suspended from the 50-ton crane. The brick building behind is the transformer house. In the foreground is the fume-extraction plant. *Courtesy of K. E. Royall*

Extracts from my [Ken Royall's] 1951 Diary

Tuesday 19 June

Work on North Pit Ladle completed. Lit the Selas burner for the first time
& it worked perfectly. Let it run through the afternoon & evening at 75"
gas & 3.5" air. Burner turned off at 9-00pm. Trial probably tomorrow.

Weds, 20 June

No test on North Pit Ladle due to Furnaces being unsuitable.

Thurs. 21 June

Informed Bill Cousins how to light the Selas burner, as the Ladle trial is
to be held tomorrow & the Burner needs to be lit at 6-00am

Friday 22 June

Ran the test on the 15-ton Ladle at 11-30am with a Heat from "E" Furnace.
Test very successful, the metal dropping only 30 degrees in 45 minutes.
Had the report typed out & sent to Capt. Baily & Mr. J.Lyon.

Tuesday 31 July

Announced that the M.O.S. was to become an "Experimental Works" &
would be managed by The Iron & Steel Board..

Transcript from Ken Royall's 1951 diary. *Courtesy of K. E. Royall.*

A reunion at the Abbey House Hotel, Barrow in 2018. Left, Gordon Jones, Fuel Technician, with authors Ken Royall (Fuel and Instrument Department) and Stan Henderson (Test House). *Photo courtesy of Trish Jones.*

– Thoughts on the term 'Invented By' –

During the research for our 2015 book – *Barrow Steelworks: An illustrated History*, Ken Royall and I soon began to realise that 'invented by' was quite often a *non sequitur* and that iron and steel making processes had been more an evolutionary thing with numerous individuals making contributions, some of which were patented.

Furthermore, we found that quite often whilst one aspect was being 'invented' say, in the UK, the same thing was occurring contemporaneously elsewhere! There are several examples of this to be found. Perhaps the best example of this co-incidence is concerning the Bessemer process.

At the time when Sir Henry was 'inventing' his revolutionary process for converting iron into steel (he was actually only trying to make a better quality of iron for use in ordnance), across the Atlantic Ocean a guy called *William Kelly* (1811-1888), was doing the same thing. The difference was that Sir Henry probably got to the patent office first. Many American text books on steelmaking, in referencing the process, place *Bessemer* in inverted commas.

Blast Furnace

It is not known who 'invented' the bf. Arriving in Britain from the Continent (probably Belgium) sometime in the 15th century it had been progressively improved upon over the centuries by numerous individuals (including our own Josiah Smith, first manager of the Barrow Haematite

Steel Company, who patented a device for capturing the hot gases that were generated by the smelting process). These early furnaces, constructed into, or adjacent to, a hillside to facilitate charging and sited next to a stream, bore no resemblance to the majestic, lofty structure of the later models. Which were then seized upon by our American cousins who devised a system of mechanical charging and thereby 'inventing' the modern blast furnace *(The first being named Eliza).*

Basic Steelmaking

The invention of the Basic method of steelmaking, (one of the several factors responsible for Barrow works losing it's almost monopoly position), is credited to *Sidney Gilchrist Thomas* (1850-1885) at Blaenavon Ironworks in Wales. Gilchrist found a way of eliminating phosphorus in the Converter. This allowed steel to be made from inferior ores containing high levels of phosphorus. Interestingly, around the same time at Workington in Cumberland, *George James Snellus* [1837-1906], who was works manager of the West Cumberland Haematite Iron Company had also found the answer. Snellus was a distinguished metallurgist and chemist. His greatest achievement was his 'invention' of basic furnace linings which he patented in 1872.

Stainless Steel

History records that metallurgist, *Harry Brearly* (1871-1948) invented a type of stainless steel (*aka* stay-brite steel), at Sheffield in 1913. However, during research for our 2015 book it emerged that a type of stainless steel was being made at Barrow *before* it was 'invented' by *Brearly*! The difference was that at Barrow it was called *chrome-steel*. But not being substantiated *via* written record, we could not include this in our book.[1]

---

1 personal communication author/ Bill Pearson, Senior Melter, BSW. (1913 to 1963).

Continuous Casting

When the success of the process development broke and became headlines in the local press, many Barrow folk, including some works personnel, were under the misapprehension that continuous casting was *Iain Halliday's* invention. In reality he had only perfected, or fine-tuned, the non-harmonic, oscillating mould technique of Herr Junghans.

And Finally

I will always remember what I was told during one of the many interviews we conducted while researching our 2015 book: It was always the done thing in British industry, where new technology was to be introduced into the workplace, to involve the Union – to give the undertaking a semblance of inclusivity or to appear democratic. On one such occasion, when Barrow works were looking at upgrading No 3 hoop mill to automatic, works convenor, Bill Miller, senior hand on the mill, was taken along with engineering personnel, to the Stocksbridge works in Sheffield to see their automatic strip mill working. While there Bill was taken aside by one of their senior hands, who said, "It's generally regarded that Sheffield is the leader with regard to steelmaking. Not so, Barrow has that distinction".

- ACKNOWLEDGEMENTS -

The authors would acknowledge the efforts of Manfred Rasch of the ThyssenKrupp organisation for his exemplary work in putting together the book – *Stranggießen–Continuous Casting* (2018) and for allowing us to contribute to this, the most comprehensive document on the subject. Thanks are also extended to Mr W. Whiteside Jnr., Frank Pearson Jnr. and Peter Caldwell, for information and images concerning their fathers.

Thanks are also due to many ex-BSW personnel for their valued contributions, including: Rodger Bradley; Bernard Hogg; Ray Millard; the late Jack Smithson and Billy Fell.

Brian Cubbon (*The Red Earth Revisited*), has been an ongoing source of encouragement and suggestion as well as valued opinion.

We cannot overstate our gratitude to the curator and archivist – Susan Benson – of Cumbria Archives & Local Studies Centre (Barrow) and also to Russell Holden of Pixel-Tweaks for producing the book.

-oOo-

- THE AUTHOR -

Stan Henderson is a Barrovian Senior Citizen, born in 1949. His industrial life (30-years) was spent largely in the drawing offices of Vickers Armstrong (Shipbuilders) Ltd., (later VSEL), where he held a management grade senior staff position. For several years during the 1970s Stan was a part-time lecturer at Barrow College of Further Education. In 1995 he took voluntary redundancy, leaving the shipyard to run a large convenience store with his sons on Walney Island. Since attaining retirement age he has co-authored two books on Barrow Steelworks. He has also penned a book about his shipyard apprenticeship as well as two local histories about Hindpool, Barrow-in-Furness. Stan now spends most of his time at a holiday home in Kirkby-in-Furness, South Lakeland.

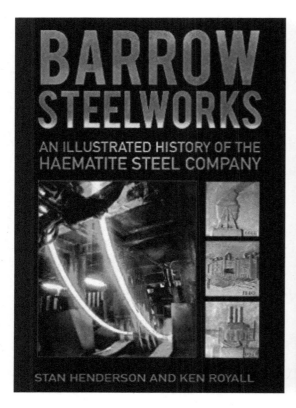

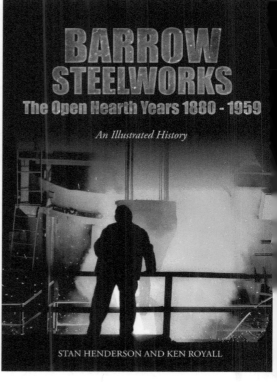

During the second half of the nineteenth century, Barrow-in-Furness became a pioneer in iron and steel production. It went on to grow astronomically – owning collieries in three counties and ore mines in two – and became the largest integrated steelworks in north Lancashire and Cumberland and, at one time, the largest steelworks in the world. Its success was due, in part, to having the prestige of three dukes as directors, as well as to being only 2 miles away from one of the largest and richest iron ore mines in the country.

The 1880s were a decade of change for Barrow works with some of the main players departing the scene. The arrival of the basic method of steelmaking, took away the lucrative position held by the directors and shareholders who had drained the coffers leaving virtually nothing for re-investment. After the Great War the company was limping along. The evacuation of Dunkirk at the start of WWII together with the blocking of special steels produced a demand for the kind of steel the making of which Barrow was a past master. Under United Steel's banner Barrow would see security of employment.

Paperback: 160 pages
Publisher: The History Press;
Language: English
ISBN-13: 978-0750963787

available at
amazon

Paperback: 98 pages
Publisher: Stanley Henderson
Language: English
ISBN-13: 978-0995619050

available at
amazon

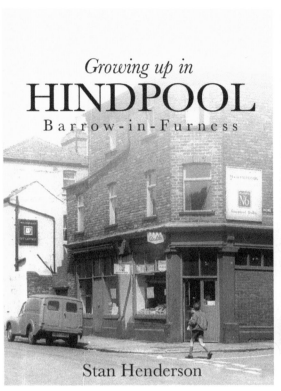

Growing up in
HINDPOOL
B a r r o w - i n - F u r n e s s

Stan Henderson

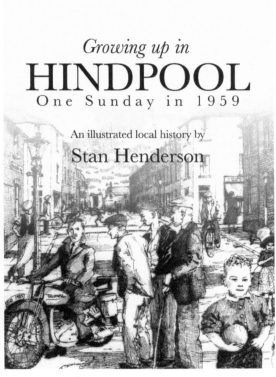

Growing up in
HINDPOOL
O n e   S u n d a y   i n   1 9 5 9

An illustrated local history by
Stan Henderson

The history of Hindpool has been likened to a patchwork quilt, with each fragment, or patch, different in time, size, shape and colour. In his book the author has woven his quilt with the thread of family history and personal experience. The story starts with the arrival in Barrow of the writer's ancestors, immigrants from Shropshire, who had come to work on the blast furnaces of the local, monster, Ironworks. These works would later hold an unexplained fascination for the author, who, in this book takes the reader on a conducted tour around the historic works.

In this follow-up to Growing Up in Hindpool, the author completes his patchwork quilt with respect to the industries, institutions and businesses to which he has been directly or indirectly involved. The reader is taken on a walk out of the district and, via Lower Cocken, into Ormsgill, then back into Hindpool. During this walk, which 60-years ago, was undertaken at least once per week, the author reflects upon aspects of 1950's life, bygone industries, landmarks and some of the local characters that made Hindpool one of Barrow's most fascinating places in which to belong.

Paperback: 140 pages
Publisher: Stanley Henderson
Language: English
ISBN-13: 978-1916021747

available at
amazon

Paperback: 84 pages
Publisher: Stanley Henderson
Language: English
ISBN-13: 978-1916275836

available at
amazon

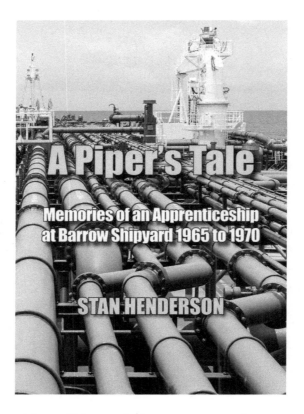

A Pipers Tale records the impressions made on a teenager as he makes his way into the thorny world of shipbuilding. A world in which the author, during the 1960s, witnessed the change from traditional shipbuilding, where vessels were constructed with a minimum, but adequate, level of technical support via long established trade practices and skills, to the cutting-edge of science-based projects as the Yard at Barrow became a 'Leader in Marine Technology' with the making of sophisticated warships and first-of-class vessels. Saluting the the wealth of characters and personalities that comprised the Yard's Plumbing Fraternity.

Paperback: 96 pages
Publisher: Stanley Henderson
Language: English
ISBN-13: 978-0995619081

available at
amazon

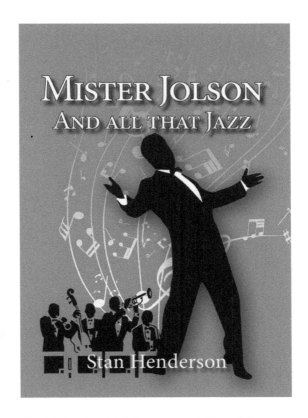

In this book, which is an appreciation of the popular art, the author takes us on a fleeting excursion through the evolution of 'pop' - from ragtime to hot jazz and in to the swing era - a fascinating insight into the early Broadway Musical and the birth of the 'talkies'.

The emergence of the Great American Song Book, and the influence of Al Jolson's career on popular singing; his relationship with the principal song writers, and how he inspired the great vocal stars who followed, including Ethel Waters, Bing Crosby, Judy Garland and Frank Sinatra.

Paperback: 92 pages
Publisher: Stanley Henderson
Language: English
ISBN-13: 978-1913898045

available at
amazon

Milton Keynes UK
Ingram Content Group UK Ltd.
UKHW051037140923
428659UK00004B/9